CAR MODELS THAT ZOOM
Creativity in Motion

ED SOBEY, Ph.D.

To Edwin and Rose Sobey

Photographs: Ed Sobey

Project editor: Mary Anne Dane

Production editor: Carol Hiltner, CarolHiltner.com

Cover design: Peter O'Connor, BespokeBookCovers.com

© 2013 by Ed Sobey

All rights reserved

First edition

Acknowledgments

I thank all of the teachers, science center staff, and students who attended our workshops and have inspired me with their creativity. Just when I think I have seen all the possible solutions to a design problem, someone comes up with a new design.

I also thank my wife, Barbara, for her assistance and patience. I started writing this while we were on a workshop tour in Austria in 2013 and jumped back to the keyboard at every opportunity.

Sometimes you just luck out meeting the right person at the right time. It was my good fortune to meet publications expert Mary Anne Dane. She guided me through the many steps needed to produce this book. Thank you very much, Mary Anne.

Several people helped by reviewing the manuscript and offering suggestions for improvements. Dr. Sally Montgomery, former science center director, innovator, and creative spirit from Northern Ireland provided great feedback in reviewing the manuscript. Denham Dunstall, chief wizard for the Perth science center, SciTech, provided good insight from down under. Ryan Collay suggested some additional activities to add. He directs the aptly named SMILE program at Oregon State University. Twelve-year-old Townsend gave me the kids' perspective and pointed out terms that kids might not know. Lori Fares, a five-star homeschool mom, and good friend from Semester at Sea, also contributed several improvements. Henry White fell off his bicycle and broke his elbow when I asked him to review this text. The elbow is healing, as is the text. My older genius son Woody provided suggestions and corrected a few mistakes. Thank you all.

National Science Network staff makes Zoom models in Vienna.

Contents

Introduction 6

Get Started 7

Materials 8
- Cardboard
- Straws
- Wood dowels
- Wheels
- Craft sticks
- Electric motors
- Wires
- Batteries
- Switches
- Balloons
- Other materials

Tools 16
- Hot glue guns
- Scissors
- Knife
- Drills
- Cutters

Models
1. Rolling car 18
2. Three-wheel car 25
3. Steerable car 28
4. Sail car 31
5. Puff car 34
6. Jet-powered car 36

7. Rubber band car....................40
8. Rubber band propeller car..........44
9. Electricity primer...................48
10. Electric car with propeller drive......54
11. Electric car with direct drive.........61
12. Electric car with gear drive..........66
13. Electric car with belt drive...........71
14. Solar-powered car...................75
15. Jittering car 82
16. Hovercraft.........................87
17. Motorcycle.........................92
18. Cable car 99
19. Radio control car..................105

Introduction

What I cannot create, I do not understand.
Richard Feynman, Ph.D., Nobel Laureate

This book is about having fun building things, messing around, asking questions, and seeking solutions. It's about making mistakes, stealing ideas, and creating understanding as well as models.

We hope that building the models we show will inspire you to create new models. We provide instructions to get you started and also solve some of the vexing problems, but we want you to use the materials you have available and generate your own design ideas. We want to jump-start your creativity.

Building when you have a goal in mind, without rigid instructions, is called prototyping. Fast prototyping is the process of swiftly trying ideas and rejecting the ones that don't work in a constant effort to make something work well. Fast prototyping lets you create models quickly and generates learning quickly. It is the fast paced way to learn.

We developed the approach of learning by fast prototyping from listening to the world's great inventors. I had the unique opportunity to talk to the greatest living inventors and learn what made them successful. As founding director of the National Inventors Hall of Fame, I interviewed inventors when they came to be inducted into the Hall of Fame. What I learned changed my life.

James Hillier, inventor of the lens for the electron microscope and holder of hundreds of patents, stated it most succinctly. "We [the inductees in the National Inventors Hall of Fame] all have one common experience in life," he told me. "We all had the opportunity to mess around in a shop or laboratory early in our careers. We had the opportunity to make mistakes when a mistake did not end a career." In messing around and making mistakes, great inventors develop confidence in their ability to solve technical problems in addition to improving their technical skills.

This book is about learning from mistakes. It's about stimulating your curiosity and creativity. Have fun making models!

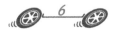

Get started

Build it; don't plan it. When in doubt, start to build. Let your hands and mind work together to solve the challenges in this book. Don't force your hands to sit idle while the mind struggles with three-dimensional thinking.

Focus on making each model as fast as you can and then think about how it performed. Expect failure and bounce back to try again. Delight in failure because, as Edison said, each failure is a stepping-stone to success and understanding.

Better to make ten mistakes an hour and have three successes than to make no mistakes and have one success.

Revel in the different ways of doing things. Look around for inspiration. Be on the lookout for a great idea you can borrow. There are wonderful design ideas everywhere.

Collect the parts you need and keep them handy. Grab the glue gun and tape. Let's start now—not tomorrow, now.

Materials

Many of the materials you need to build car models you have already. Look around the house to see what you can substitute for any of these materials.

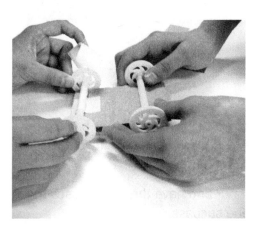

Cardboard

A cardboard box provides material to make car bodies. It has other uses as well. If you intercept it on its way to being recycled, it is free.

Cardboard comes in many forms. For most models corrugated cardboard works best. But, save those empty cereal boxes for their single-ply cardboard. It is a great material for many uses.

Straws

You will need to attach the axles to car frames so the axles can spin. How do you stick something to the cardboard and yet allow it to turn? You use a bearing. Straws work well as bearings. Find straws that are just a bit larger than the axles you are using.

Wood dowels

Wood axles are ideal. They are easy to cut and are inexpensive. Ideally they fit into the center hole of the wheels you select. Check the sizes of both wheels and axles before purchasing either. Hardware and craft stores sell "dowels." Grocery stores sell wood sticks to use as skewers in cooking. We use dowels that are 1/8 inch or approximately 3 mm in diameter and occasionally use dowels twice that size (1/4 inch).

> Materials

Wheels

Plastic wheels are great, but there are many other options. For example, you could use old CDs or you could cut your own wheels out of wood, cardboard, or plastic.

But if you want to get your car rolling fast, consider purchasing wheels. For most models we use 1 inch (2.54 cm) wheels with 1/8 inch (3.175 mm) center hole.

Plastic wheels sold by science stores online

To cut wheels out of plastic sheets or wood boards, purchase a hole saw. They come in a variety of sizes. Insert a hole saw into the chuck of an electric drill and you are ready to make wheels.

Wood wheels cut with a hole saw (right)

Wheels that have gear teeth cut into the rim allow you to drive the wheel using a gear on the motor shaft. Check the diameter of the center hole in the gear wheels to make sure your axles will fit.

Gear that mounts on a motor shaft (left) and gear wheel

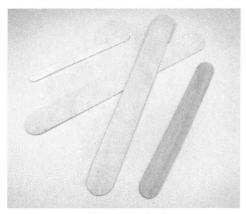

Craft sticks

Craft sticks

These wood sticks come in a variety of sizes. They are useful for many projects and are inexpensive, especially if purchased in a box or large bag. Keep a box or bag of these handy on your workbench.

Electric motors

Toy motor (left) and geared motor

Low voltage or solar motors

Motors come in a wide variety of sizes, strengths, and speeds. For some projects a low torque (turning power), fast spinning motor is best. These are called toy motors and you can collect them from broken toys. Or, you can buy them from science stores. They are inexpensive so you can purchase several.

Some projects require motors with a lower rate of spin. Consider the different requirements. To propel a model car with a propeller, the motor must spin fast: for example, 15,000 RPMs (revolutions per minute).

> *Materials*

To power a car with a gear or belt, 15,000 RPMs would be too fast and might not deliver enough torque (turning power) to spin the wheels. So for this application a more expensive, slower turning and higher torque motor would be better.

Some motors require 3 or more volts. These work well with batteries, but are not as useful for solar-powered projects.

Gear motors have a series of gears built into the head of the motor. You can't see the gears because they are enclosed. The motor itself spins fast, but the gearing reduces the output shaft speed to a slower rate: 100 RPM for example. These motors would never spin a propeller fast enough to move a model, but can push even a heavy car with gear drive.

When powering models with solar cells, motors requiring lower voltage work better than the inexpensive toy motor. Solar cells output about 1.5 volts and toy motors typically operate at higher voltages. Of course you can connect several solar cells in series to get higher voltages. We will show examples in the solar-powered car description.

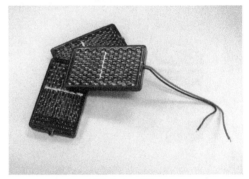

Solar cells

Another consideration in purchasing motors is the shaft size. You will save time and effort if the motor shaft fits any gears or pulleys you will want to put on it. If not, you may be able to purchase bushings or make bushings (devices that connect parts that have different shaft sizes).

Cars Models that Zoom > > > > > > > > > > > > > > > > > > >

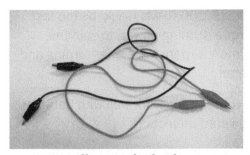

Alligator clip leads

Wires

To get electricity from the power source (usually batteries or solar cells) to the motor, you need conductors. Alligator clip leads make the process of connecting the components quick and easy. After making prototypes if you want to make a nice model that you won't change, use electrical wire (gauge 22-24) that you can cut to the exact length you need. But for fast prototyping clip leads let you build as fast as you think.

Alkaline transistor or 9-volt battery (left) and AA batteries

Batteries

If you're going to be making many models, consider investing in rechargeable batteries and a charger. Reusing batteries is good for the environment and for saving money, but only if you will charge them as they need to be charged.

Most likely you will use NiCad batteries. These batteries use nickel and cadmium. They suffer from memory effect: they lose capacity if they aren't fully discharged before recharging. That means you have to drain the battery (connecting it to a motor until the motor stops spinning, which is a waste of energy) before recharging it.

Inside a transistor battery

> *Materials*

Alkaline batteries deliver more power for longer periods of time than do most other batteries, but you have to recycle them after using them. Do not throw them in the trash. And, all alkaline batteries are not created equally. Cheap ones last only minutes. Compare the specifications on two different brands of batteries to see which provides more power. Power is specified in amp-hours or how many hours the battery can supply one ampere of current.

Transistor or 9-volt batteries deliver lots of power, but are consumed in a short period of time. Open up a spent battery to see that inside are six very small battery cells. Each cell produces 1.5 volts. The chemical reaction that transforms chemicals stored inside the battery into electrical power produces 1.5 volts. To get 9 volts from one battery manufacturers connect 6 tiny cells in series (positive terminal of one connects to the negative terminal of the adjacent cell). These tiny cells don't last long as they don't have much volume of chemicals.

As handy as 9-volt batteries are, they are the most expensive form of electricity you can buy. Some applications, like the hovercraft, require the higher voltage and low weight that the transistor battery delivers.

For most projects, however, AA batteries are a better choice. They cost less and last longer than transistor batteries. To use AAs you need a battery case. These come in a variety of sizes: to hold one battery, two batteries, or four batteries. The battery holders are inexpensive so consider purchasing one of each size.

Battery holder for AA batteries

Switches

Although they are not necessary, switches make it easier to operate motors. Rather than trying to connect clip leads to battery terminals while holding the model in an awkward position, you just push the switch closed.

Knife blade switches

If you purchase switches, you probably want a knife-blade switch. You push the blade down to close the circuit and let electrical current flow from one side of the switch to the other. This type of switch stays open or stays closed until you lift or lower the knife blade.

Momentary contact switch

Momentary contact switches are open only as long as you depress the lever. These also work in the "normally closed" position so they complete the circuit unless the lever is depressed.

Switch made from paper clips

Of course you can make your own. Here is one example that uses brass clips (from an office supply store—these hold paper pages together) and a steel paper clip. Slide the paper clip so it makes contact with the second brass clip to complete the circuit. Clip leads can attach to the legs of the brass clips.

Maybe the best option for switches is to collect them from things you take apart.

Home and office appliances and electric toys have switches that you can remove and use. Before discarding a broken appliance, take it apart to get the switches.

Switch made from paper clips

For a guide to taking apart appliances, see my book, *Unscrewed: Freeing Motors, Gears, Switches, Speakers, and More from Your Old Electronics*.

Balloons

Large, round, latex balloons are best for models. We use 12 inch or 30.5 cm diameter balloons. Try other sizes or shapes, but we like the round ones.

Other materials

Rubber bands—a variety of sizes

Tape—masking tape, not clear tape and not duct tape. Masking tape is inexpensive and easy to tear with your fingers. The best size is 2 cm wide.

String—light weight nylon string

Paper or polystyrene dinner plates

Solar cells, super capacitors and voltmeters—see the specifications in the section on solar-powered cars.

Tools

Hot glue gun

Glue guns are great. They allow you to build as fast as you think. The glue holds well—not when it's wet—but can usually be pulled off if you make a mistake. And plastic glue is made for mistakes. It's not permanent glue; it's fast prototyping glue.

Hot glue gun

Glue guns come in several sizes. We recommend using the smaller size glue guns and sticks (1/4 inch). The smaller size has fewer features and costs less. These guns can operate for years.

If you ever have a broken glue gun, take it apart. First make sure it is not plugged in. Then cut off the electrical plug so it won't get inserted into an outlet while you are exploring the insides. Two or three screws hold the glue gun together. Separate the two halves and marvel at the simplicity of the design.

Glue guns burn skin, so be careful. If you get molten glue on your hand hold it under running water to relieve the pain.

Scissors

Scissors are a must for cutting cardboard and paper—the materials of many fast prototyping projects. Have different sizes available.

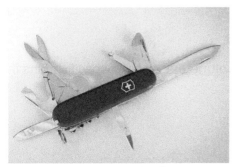
Swiss Army knife

Knife

For making long cuts in cardboard and for a thousand other uses, a Swiss Army knife is tops. Other knives will do, but if you carry a Swiss Army knife in your pocket you will always have the tools you need for most jobs.

Drills

Electric drills are useful, but a less expensive option is to purchase only the drill bits you need. You can mount each one in a handle and turn the drill bit by hand.

What size drill bits do you need? The most useful sizes are the diameter of the center hole in the wheel you are using and a slightly larger diameter drill that an axle can fit through.

Hand turned drills

Use a drill press or electric drill and your new drill bit to drill a hole in a dowel or some other type of handle. Glue the drill bit in the handle. You are ready to turn.

Cutters

For cutting small dowels (1/8 inch or 3 mm) and wires a diagonal cutter will do the job. For larger dowels consider a saw with fine teeth. Or, purchase one hacksaw blade and build a handle for it from wood and duct tape.

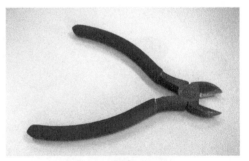

Diagonal cutters

Safety

All of these tools are safe when handled properly and all can cause injuries if not handled properly. If you aren't sure how to use a tool, get some expert help at the outset.

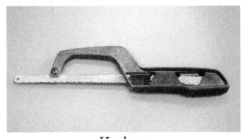

Hacksaw

① A rolling car

Build a car that can roll down a ramp or be pushed across the floor. The goal is to get the model to roll straight and far.

Materials

- Cardboard—a few small pieces of corrugated cardboard
- One straw
- A dowel (1/8 inch or 3 mm diameter)
- Wheels (4) that fit onto the dowel
- Masking tape or hot glue
- Scissors
- Diagonal cutters

Build the model

Cut a rectangle of cardboard about 8 cm by 12 cm (3 by 5 inches).

Glue or tape two sections of straw to the bottom of the cardboard. To have the car go straight make sure the two pieces of straw are parallel to each other. It is best to place them near the ends of the cardboard.

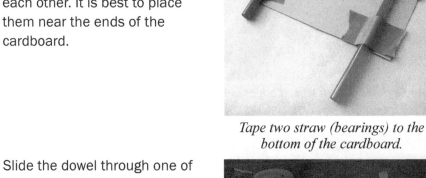

Tape two straw (bearings) to the bottom of the cardboard.

Slide the dowel through one of the straws and use the cutters to cut the dowel so it protrudes slightly from each end of the straw.

Attach wheels to each end. Now the dowel is an axle and the straw is a bearing that holds the axle in place. Repeat this step for the second dowel.

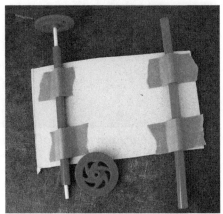

Slide an axle through each bearing and add wheels.

If the car turns, align the axles. Make the distance between the axles the same.

A wood plank with raised edges is the ramp in these experiments.

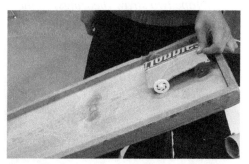

Launch the car from the top of the ramp to get the farthest roll.

Test the model

Roll the car down a ramp. You could use a large piece of cardboard resting on stairs for a ramp. Or, collapse the legs on one end of a table so its surface becomes a ramp. Start the car at the high end of the ramp and watch it roll down.

Do the wheels turn as the car goes down the ramp? If the wheels don't spin easily, the car will turn to one side.

Do the wheels rub on the cardboard or on the straw bearing? If so, the car will quickly stop.

Does the car turn to one side? Test the car backwards to see if it turns to the other side. If it does the axles are probably not parallel to each other.

Watch the car at the bottom of the ramp. Does it jump when it gets to the end of the ramp? There are several things that could make it jump. One is that the car frame, the piece of cardboard, hits either the ramp or the floor. The wheels aren't high enough or aren't close enough to the ends of the cardboard. Hold the car at the bottom of the ramp and roll it by hand on and off the ramp to see if part of the car hits the ramp or floor.

Another possible cause for the car jumping is that it is top-heavy or that the weight of the car is not centered.

Measure and improve

If you are using a ramp to test your car, measure how far the car rolls across the floor. Measure the distance it goes on each test. Measure from the base of the ramp in a straight line.

What can you modify to make the car roll farther? Measure each test and record the distance. Make one change each time you test the car. By making one change at a time you will learn the impact of that change on the car's performance.

Science project

A simple science project is to measure how far the car travels and retest it after making one change. Start by releasing the model from the top of the ramp and then make a change. Measure how far it goes. Then make another change and retest it. Record the distances and the change you make each time.

Instead of wheels, this car rides on practice golf balls.

You could start by adding weight to the car. After you test the car, add more weight and measure how far the car goes.

Create a graph to show the weight you added for each test and the distance the car traveled. Each test gives you a pair of data (weight added and distance traveled) that you can graph. The graph tells a story about energy: the more potential energy (weight at the top of the ramp) you put in, the more kinetic energy (speed of the car at the bottom of the ramp and hence the distance it travels across the floor) you get out.

Another science project

Once you have a car that works well, measure the height of the top of the ramp and measure how far the car goes when released from the top. Record these two measurements as one data pair. Then repeat the experiment at a slightly lower elevation on the ramp, measuring how high above the floor the car starts. Keep repeating the experiment measuring both the height of the car on the ramp where you release it and the distance from the bottom of the ramp to where the car stops.

The final data pair is at no elevation (0 cm above the floor) and no distance traveled (0 cm). Graph the pairs of data on a piece of graph paper. The graph tells a story that is fundamental to understanding the universe: the more energy you put into a system, the more you can expect to get out.

F *represents force;* **m** *represents mass or weight of the model, and* **a** *represents acceleration. If you increase the force and the mass is constant, the acceleration increases.*

Science concepts

The force of gravity pulls the car down the ramp. The greater the force, the faster it can accelerate the car.

Gravitational force is controlled by the acceleration of gravity—something you cannot change—and the mass or weight of the car. You can increase the mass of the car and increase the force of gravity pulling the car down the ramp. The more mass (weight) you add, the faster the car will go and the farther it will go.

This increase occurs only up to a point. When the weight starts to bend the axles or cause the wheels to rub on the car frame, the distance will not increase.

Force and energy are different concepts in science. Sitting at the top of the ramp your car has gravitational potential energy. It has the potential to do work: release the car and it will move.

Where did this energy come from? You supplied the energy when you lifted the car to the top of the ramp.

Would a six-wheel car go farther than a four-wheel car?

Where did you get this energy? From the bowl of cereal you ate for breakfast. The cereal got its energy from plants that converted the sun's energy into sugars. So your car is a solar-powered car. But so is just about everything else that moves, lights up, or makes noise.

The sun is the ultimate source of energy for almost everything on our planet.

The car had potential energy at the top of the ramp and when you released the car it moved. The potential energy was converted into energy of motion or kinetic energy.

But if the car had kinetic energy, why did it stop? Something had to decrease the amount of energy the car had. That something is friction.

Two film canisters filled with water provide weight for this model.

Cars Models that Zoom > > > > > > > > > > > > > > > > > > >

As the car rolled down the ramp, the axles rubbed against the bearings and the wheels rubbed against the ramp. This rubbing transforms the energy of motion into heat, yet another form of energy.

If we were able to roll a million cars down the ramp very quickly the room would get hot from the heat generated by all the friction. In a tiny way your gravity car is a solar-powered heater.

Making the car heavier often pushes it farther across the floor. Adding weight increases the cars mass and momentum. Momentum is mass multiplied by velocity.

When the car moves from the inclined ramp to the floor, the more momentum it has the farther it will roll. But as you add more weight the axles bend more and rub against the bearings. The car frame might bend as well. So there is an optimal weight that propels the car farthest.

A car with three wheels

Here's a model powered by gravity, but one with only three wheels. It's easy to build a car with four wheels, but how will you support one wheel without another wheel on the opposite end of an axle?

Materials

- Cardboard—a piece about 8 cm by 12 cm (3 by 5 inches) or craft sticks
- A straw
- Wheels (3) that fit the dowel/axle
- An axle or dowel (1/8 inch or 3 mm diameter)
- Masking tape or hot glue
- Scissors

Build the model

Attach one axle with two wheels as in the first project. For the second axle, the one with only one wheel, force a short (6 cm or 2.5 inch) piece of dowel through the center hole of a wheel. Position the wheel in the center of the dowel.

Cut two short sections of straw, each about 3 cm long. Put one piece of straw on each end of the dowel. Glue or tape the two straws to the bottom of the cardboard or craft stick frame.

Using a short axle makes it more difficult to get the two axles aligned.

Attach one axle with two wheels.

This makes it more difficult to get the car to go in a straight line.

To align the two axles, draw a line on the bottom of the car to show where the second axle should be. Make this line parallel to the first axle by measuring the distance on each side and marking the same distance. Draw a straight line through the two marks you made.

Test the model

Use the ramp from the first project to test the three-wheel car. Check to see if the third wheel is rubbing against either piece of straw or the cardboard frame. If the car turns to one side the axles are probably not aligned with each other.

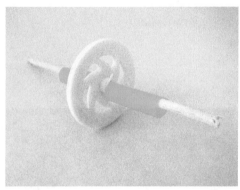

The single wheel is mounted in the center of its axle and is supported on each side with a straw bearing.

Roll the car backwards down the ramp to see if it turns to the opposite side.

> > > > > > > > > > > > > > > > > > > A car with three wheels

If you used the same car frame for the three-wheel model as you did for the four-wheel model, which do you think will roll farther across the floor? Test the new model and record the distance it travels.

Things to look for

See if the single wheel rubs against the bearings or the car frame. If it does, this will slow the car.

Single wheel is taped or glued in place.

Here is a three-wheel car built with craft sticks.

③
Steerable car

Can you figure out how to have the car turn to the left or right? And, can you make the direction of the turn and the angle of the turn adjustable?

Materials

- ▶ The model you built for either Project 1 or 2
- ▶ Scissors
- ▶ Large paper clips or binder clips, or
- ▶ Brass paper fasteners

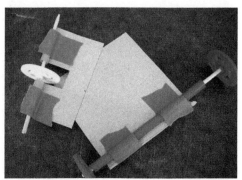

The axle is rotated so the car will turn.

Build the model

Angling the front axle will cause the car to turn. If you want the direction of travel to be adjustable, you need a way to hold the axle in position until you want to change the angle.

> *Steerable car*

A simple way to make the car's steering adjustable is to cut off the front axle. Use scissors to cut across the front of the car leaving about 2 cm (3/4 inch) of cardboard. Reattach the axle by holding it in place with either a binder or paper fasteners. You can rotate the axle to the angle you want and then clip it in place.

A "bulldog clip" holds the front of the car onto the frame so the steering angle can be adjusted.

Test the model

Try several different angles to see if your steering works well. Check to see if the wheels rub against the car frame when the steering angle is large.

Try this

A more elegant way to make the steering adjustable is to use a brass paper fastener. Punch a hole through both the cardboard piece holding the steering axle and the cardboard frame. Slide the brass fastener through the two holes and spread the two legs to keep it in place.

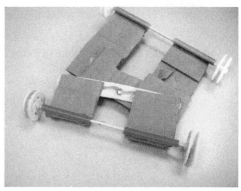

A brass paper fastener holds the front axle assembly to the car frame.

You could easily replace the brass fastener with a small nut and bolt.

Brass paper fasteners

A wood model has adjustable steering axle held by a screw.

Science concepts

Why does the car turn? With the wheels turned to one side the car would have to overcome considerable friction to travel straight: the wheels would have to drag across the ground.

Less energy is required for the wheels to rotate around the axle rather than drag on the ground so the car goes in that direction.

Sail car

An easy way to power your three-wheel or four-wheel car is with a sail. Take a look at pictures of sailboats to get some ideas for designs.

Materials

- ► Your model from Project 1 or 2
- ► String
- ► Paper and plastic scraps, or
- ► Cloth scraps
- ► Masking tape
- ► Cereal box
- ► Hot glue
- ► Dowel or small dead branch from a tree
- ► An electric fan

Build the model

If the wind is reliably blowing from behind your car, a square sail will work well. Cut a piece of scrap paper, plastic, or cloth into the shape you think will capture the most wind.

Cut a dowel or straight tree branch that is shorter than the length of the car. How will you keep this mast upright? You can start by gluing or taping the base of the mast in place, but that won't hold it upright. As soon as the wind blows on the sail, the mast will fall.

To support the mast you could use either string or pieces of cardboard glued to the car frame and mast. Once you have the mast in place, attach the sail, and test it.

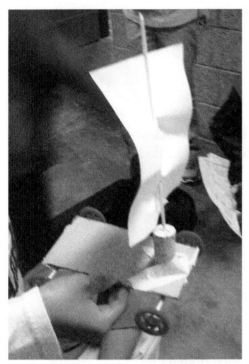

Sail car with a square sail

Testing how far the sail car will travel when powered by a fan

But the wind isn't always blowing from behind the car, so how can you sail the car if the wind is coming from the side? Instead of using a square sail, use a triangular sail. Glue the leading edge to the mast. Use a piece of string to hold the back end of the sail in place. You might add a boom (a piece of dowel) to the bottom of the sail to help hold it.

Test the model

Use the wind if possible. If it isn't blowing use a fan or hair dryer. Can you get your car to travel several meters? Can you get it to sail when the wind is coming from one side? Adjust the angle of the sail to get the best movement when the wind is not behind the car.

Consider this

What if you mounted a fan on the car so it could blow onto the sail? Would that work? Think about that and you can try it in the propeller car model.

Science concepts

It took sailors thousands of years to figure out how to use sails to push boats. Picture the ships in *Pirates of the Caribbean*. It took even longer to figure out how to use sails so the boats didn't have to sail downwind—so they could travel almost directly into the wind. Look at the sailboats you see at a marina. See if you can figure it out how to get your sail car to sail closer to the fan.

Sail car with triangular sail cut from a cereal box

Puff car

Lung power propels this model. You exhale as quickly as you can into a bendable straw and that pushes this model across the floor.

Materials

- Wheels—plastic wheels or CDs
- Straw—bendable
- Straw—large diameter straw for milkshakes, not bendable
- Dowel
- Craft stick
- Hot glue

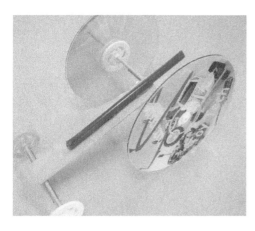

Build the model

This model needs to be light so use a single craft stick for a frame. Glue a section of straw at each end to be the front and rear bearings. The bearings need to be parallel to each other for the car to travel straight.

Insert a section of a dowel into a wheel. Slide the dowel, now an axle, into a bearing. Attach the second wheel on the other side. Repeat for the other axle and bearing.

Fill one end of a milkshake straw with hot glue. Let the glue freeze to block any air from leaving. Glue this to the craft stick so the closed end of the straw is at the front end of the model.

Test the model

Insert a bendable straw into the open end of the milkshake straw. With the model resting on a smooth floor, blow as hard as you can into the bendable straw. The car will speed away leaving you holding the bendable straw.

A craft stick frame with one set of wheels attached

Try this

Measure how far the car goes on a smooth floor. Try adding small weights to see if it improves the distance. A single AA battery would be a good weight to test. Does adding weight improve the model's performance?

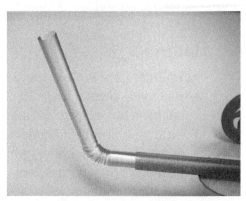

Blow hard into a bendable straw to launch the car.

Try different size wheels. If you started with 5 cm plastic wheels, glue CDs or DVDs to the plastic wheels. Make sure you have the CDs or DVDs centered on the axles.

Jet-powered car

If you don't have a jet engine handy use a balloon. You will have to exert energy to stretch the material of the balloon and inflate it. The air pressure of an inflated balloon is higher than the air pressure outside the balloon, which is why the air escapes if you release it. So the balloon stores the energy you exerted when blowing it up. That energy is stored in the elastic fabric of the balloon and in the elevated pressure of air held inside the balloon. The trick in making a balloon-powered car is figuring how to use that energy to make the car move.

Materials

- Your car from Project 1 or 2
- Balloons—several large 30 cm (12 inch) or larger round latex balloons (unless you are allergic to latex)
- Straws—a variety of diameters
- Acrylic or plastic tubing—get short lengths of various diameters from a hardware store (optional)
- Masking tape

Build the model

Inflate a balloon and release it. As it flies through the air it demonstrates that it has lots of energy you can use to propel a car. But it doesn't fly in a straight line and it empties in seconds. Overcoming these two problems is the key to getting your car to go far.

> *Jet-powered car*

When you released the balloon it flew randomly about. But by attaching it to a car that stays on the ground it travels in one direction (due to the friction of the wheels on the ground).

You need to consider where the balloon exhaust should point: up, down, to the side, or straight back.

Insert a short section of a straw or tubing into the mouth of the balloon. Wrap the neck of the balloon around the straw and tape it in place. Inflate the balloon and tape the straw to the frame of the car. In which direction will the balloon push the car?

Test the model

Release the model with the balloon inflated and observe how far it travels.

Things to look for

When you released the inflated balloon did the car move? If the balloon empties so fast that the car didn't move, consider reducing the weight of the car and slowing down the speed of the air leaving the balloon by using a smaller diameter straw.

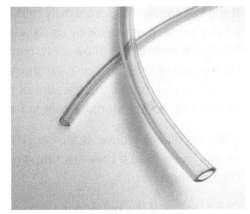

Plastic tubing is available in a variety of sizes so you can experiment to find the optimal size.

Insert a short piece of straw into the mouth of a balloon and tape it in place.

Cars Models that Zoom > > > > > > > > > > > > > > > > > >

If the balloon empties so slowly that the car doesn't move, speed up the exhaust of the air by using a larger diameter straw or tube. A larger diameter will allow air to come out of the balloon faster. Or, try cutting the straw or tube shorter. As you reduce the length of the straw you reduce the amount of drag the straw exerts on the air flowing out, so the air can exhaust faster.

If the car doesn't move at all, check that the wheels are still able to spin freely. See if they are rubbing on the frame of the car or on the bearings.

Check to see if the balloon is folding back on itself, shutting off the flow of air through its neck. The balloon can flip backward as the car starts to move, stopping air from escaping from the balloon.

Also check to see if the inflated balloon is dragging on the ground or getting caught underneath the car.

Measure how far the car travels and record the distance.

Try this

What can you do to get the car to travel farther? Consider making any of the following changes and try them one at a time, recording your results:

1. Shorten the straw sticking out of the balloon
2. Lengthen the straw sticking out of the balloon
3. Add weight to the car
4. Remove weight from the car
5. Use a larger or smaller diameter straw
6. Change the angle of the straw.

Could you add a second balloon to get the car to go farther? Would you put the second balloon beside the first one? Or, inside the first one?

> Jet-powered car

Science concepts

What did you learn from these experiments?

What is the most effective way to use a balloon to power a car?

For the car to move in one direction, the balloon has to blow in the opposite direction. The balloon forces air out and the departing air pushes the balloon (and car) in the opposite direction.

Once you have a one-balloon car that works, try making a two-balloon car.

When you tested the gravity car the more weight you added (up to a point) the farther the car traveled. As you added more weight you were increasing the force of

Balloon-powered car with a nozzle cut from the neck of another balloon

gravity pulling the car down the ramp. But in the balloon-powered car added weight doesn't increase the force available to propel the car. The balloon is supplying the force, not gravity. The more weight (mass) the balloon has to get moving, the slower the acceleration will be. This car should be as light as possible.

Jet engines in airplanes burn fuel to heat air. The hot air expands and pushes backward so the airplane moves forward. The balloon also pushes air one way so the car moves in the other direction. A balloon doesn't burn fossil fuel to heat the air. Instead you had to burn food to provide the energy to inflate the balloon.

As you blow air into the balloon the elastic of the balloon gets stretched: the balloon gets bigger. Just like stretching a rubber band the stretched balloon has potential energy. The stretched elastic pushes air out the neck of the balloon.

⑦
Rubber band car

If you stretch a rubber band between two fingers and release one end you see the energy propel the rubber band across the room. You added the energy in the rubber band when you stretched it. You can use the stored energy to drive a car.

Materials

- The model you built in Project 1 or 2
- A bag of different size rubber bands
- Balloons
- Craft sticks
- Scissors

Build the model

Craft fairs and toy stores sell wooden toy cars that are powered by rubber bands. If you can find an image of one or see one at a craft fair examine it closely to see how it works.

Although it sounds easy to propel a car with a rubber band, there are some difficulties you have to overcome.

> *Rubber band car*

You need to change your model or build a new one to start. You will want to wrap one end of a rubber band around the center of one axle and that will require you to remove parts of the car frame and of one bearing. The rubber band should not rub against the car frame or bearing.

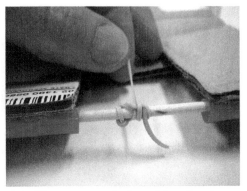

Wrap one end of a rubber band around one axle.

The other end of the rubber band has to be attached to a solid point on the frame. You could cut a small notch in the end of the frame to hold the rubber band or glue and tape a craft stick to the frame to hold the rubber band.

To operate the car, attach one end of the rubber band to the car and wrap the other end around the exposed axle. Turn the axle to roll up the rubber band over itself. When the rubber band is stretched, place the car on the floor and release your grip.

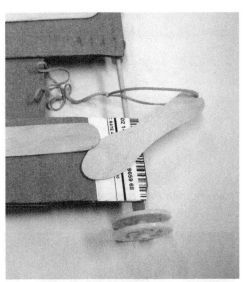

Attach the rubber band to an anchor at the other end.

Cars Models that Zoom >

Cut the necks off balloons to make tires for the wheels so they have more traction.

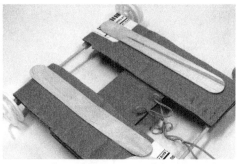

Strengthen the frame with craft sticks.

A race of rubber band cars

Things to look for

Does the car move forward? Or, do the wheels spin while the car sits still? The wheels might not have enough traction on the floor. You could increase the friction of the wheels by wrapping them in rubber. Cut the neck off of a balloon and stretch it over the two drive wheels. You are making tires for your wheels. You could also add weight directly above the drive wheels.

Does the rubber band expend its energy too quickly? Consider adding a second rubber band and lengthening the distance the rubber band pulls. You might need to glue a dowel to the frame to extend the attachment point for the rubber band.

If the car frame buckles or bends under the stress of the stretched rubber band, glue strips of heavier cardboard or woodcraft sticks to strengthen the frame.

Test the model

Measure how far your car travels. Try driving the car both as a rear-wheel drive and as a front-wheel drive. Make the switch by wrapping the rubber band around the axle in opposite directions. Does the car travel better as a front wheel drive model?

Science projects

How many rotations of the axle does it take to drive your car two meters? You could conduct a nice science experiment by measuring the distance the car goes depending on how many times you wrap the rubber band around the axle. When you have a set of eight or more data pairs (distance traveled and number of wrappings of the rubber band) you could plot the data on graph paper. What story does the graph tell?

A three-wheel rubber band car

Science concepts

For the car to move forward, the wheels have to turn in the opposite direction. Does this seem natural?

When you walk forward, your feet push backward. This is more obvious if you skate on ice or roller skate. The wheels push backward on the ground and the ground pushes the wheels forward.

The rubber band stores energy like a spring does. There are many inexpensive toy cars powered by a metal spring that is inside. Try taking one apart to see how the spring is wound up and how it releases the energy. There is more here than is first apparent. See my book, *The Way Toys Work*.

Another science project

An inexpensive but powerful spring is found in the mechanical mousetrap. Can you design a car that is powered by a mousetrap?

(8) Rubber band propeller car

This car is easier to build that the previous rubber band car, but requires one additional material: a propeller designed to be spun with rubber bands. Usually these propellers propel airplane models made of lightweight balsa wood. If you have such a plane you can borrow the propeller assembly and use it to make this car or even to power a model boat.

Materials

- The car model you made in Project 1 or 2
- A propeller assembly from a balsa wood plane (science catalogs and hobby stores sell these)
- Rubber bands
- Craft sticks, dowels, or other wood

Build the model

Hold the propeller assembly in one hand with a rubber band attached to the propeller. Stretch the rubber band to get an idea of how far away it should be anchored to get enough tension. If it is too close, it won't make many turns of the propeller. If you stretch it too far, the rubber band will break.

> > > > > > > > > > > > > > > > > > > Rubber band propeller car

Cut a piece of wood the length you think is best. Shape it so one end fits into the propeller assembly and glue it in place. Attach the other end of the rubber band to the distant end of the wood. You could cut a notch in the wood to hold the rubber band or tape a paper clip to the wood and use it to hold the rubber band.

Wind up the rubber band by spinning the propeller by hand. If you have a small electric drill you could use this to wind the propeller. You want as many turns in the rubber band as you can get without it breaking.

The more energy you store in the rubber band by twisting it, the more energy you will have to propel your car.

Shape a craft stick so it can fit into the propeller assembly.

Test the model

With the rubber band twisted as much as you think it will take, place the car on the floor, and release your grip on the propeller. How far does the car go?

If it doesn't move at all, did the propeller stop spinning after a couple of seconds? If that is the problem you need more turns on the rubber band. You might need to add a second rubber band to get more tension.

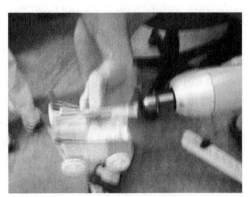

Hold the propeller against the end of an electric drill to wind up the rubber band.

Try two propellers on the car.

If the car doesn't move make sure that the wheels are free to spin. Also make sure that nothing is dragging on the floor.

Each time you test the car count how many times you turn the propeller to wind it up. Record this number and record how far the car goes.

Science project

If you do ten tests of the car counting the number of turns you give the rubber band for each test and the distance the car goes you will have enough data for a good science project. Plot the data as a graph (number of turns along the bottom of the graph and the distance the car travels along the side of the graph). Lightly connect the data points with a pencil line. The line represents what you think would happen if you had taken many more data points. The real data are the points you plotted—the line represents a best guess at reality.

What story does the graph tell? Is the line straight? A straight line would suggest that for every additional turn of the propeller you get a corresponding increase in the distance. If the line is not straight what story does it tell? If the data are scattered about the graph and don't have any consistent pattern you need to test the car again and get more consistent data.

Try this

Cut a paper milk or juice carton in half to make a boat hull or use a piece of plastic foam from the grocery store. Attach your propeller assembly to it. Wind up the propeller and watch it move across a pond or bathtub.

> > > > > > > > > > > > > > > > > > > Rubber band propeller car

Science concepts

Rubber bands and other elastic material store energy. You use your strength to force the elastic to stretch. As you pull harder or twist more, the elastic stretches more. The force is proportional to the distance the rubber band has stretched.

The propeller assembly can power a boat, too.

What happens when you reach the limits of elasticity? When the molecules of the rubber band cannot stretch any more, it breaks. Wood, plastic, and many other materials work the same way: they break when they are unable to stretch or bend any farther when a force is bending them.

Look at the propeller closely. It is not two or three pieces of plastic attached to a center hub. Each of the blades of the propeller is twisted so it pushes air to the rear. Without the twists on the propeller blade it wouldn't propel the car. Next time you see a wind turbine, look at the shape. It will be similar to the propeller.

Notice the twist in the blades of a propeller.

⑨
Electricity primer

The next models use electric motors for propulsion. If you are not familiar with electric motors and electric circuits you will find this section useful.

Car models in the projects that follow use direct current (DC) motors that run on the energy supplied by batteries. These are electrically safe: you will not get shocked when using the batteries. The voltages of the batteries are too low to force an electrical current through you—or the current that is passed through you is so low that you won't feel it. An interesting exception is your tongue. Hold the terminals of a 9-volt battery on the tip of your tongue and you will feel a mildly unpleasant tingling or shock.

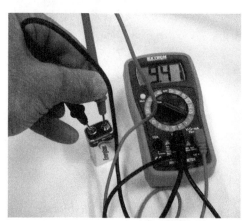

A voltmeter: connect the two probes to a battery and compare the reading to the voltage printed on the side of the battery.

Motors

Motors convert electrical energy into motion or kinetic energy. They do that by creating electromagnets inside the motor that push and pull against permanent magnets that are also inside. Electrical current flowing through a coil of wire creates an electromagnet.

Two contacts carry electrical current into and out of the motor. These connect to the positive and negative terminals of a battery. They pass the electric current to any of three coils of wire inside the motor.

> Electricity primer

Like permanent magnets, electromagnets have positive and negative poles and attract metal. Stopping the flow of current causes the electromagnet to weaken and die quickly.

Passing an electrical current through in the other direction causes the formation of another electromagnet, but with its poles in the opposite direction to the first one. By constantly switching the direction of electrical current through a coil, the polarity of the magnet flips back and forth.

Two permanent magnets inside the motor

The positive pole of an electromagnet is attracted to the negative pole of a permanent magnet. Opposite poles of each magnet attract and similar poles repel. The magic of a simple DC motor is that by spinning the motor constantly changes the direction of electrical current in the three coils inside. So they are constantly pulled towards and pushed away from the poles of the permanent magnets.

Electric contacts on the outside are part of the conductor that holds the brushes on the inside. The brushes are the small dark cubes near the center.

Commutator and the electromagnets (right)

Commutator on the left side and the three electromagnets

If a motor breaks, take it apart. By taking a motor apart you can see how this magic occurs. The electrical current is supplied to the motor through the contacts. Inside the motor, each metal contact has a tiny carbon brush. The brush is black. If you look closely you can see that it has been worn away by rubbing against the metal shaft of the motor.

On the shaft you can see black lines of carbon where the brushes rub. The metal contact that holds the brushes acts both as a conductor of electricity and as a spring to hold the brushes against the shaft.

This part of the shaft is called the commutator. It carries electrical current to each of the three coils. Looking closely at the commutator shows that it is not one continuous piece of metal, but is segmented. As it rotates inside the motor each segment rubs against one of the brushes. Thus each segmented part picks up electrical charge to energize one of the three metal coils to make an electromagnet. As the motor spins the commutator constantly changes the polarity of the electromagnets so they are constantly drawn to and pushed away from the permanent magnets.

Batteries

Where does the energy come from to spin the motor? The chemical reaction inside the battery converts the chemicals stored inside into different chemicals and releases energy in the form of electrons. The electrons flow through the conducting wires to the motor.

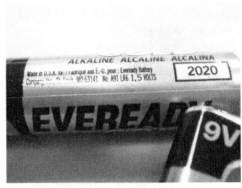

Batteries are marked with their voltages: 1.5 volts on the AA battery and 9 volts on the transistor battery.

Take a look at a battery. Notice that it is labeled to show which is the positive (+) side and which is the negative (-) side. Also see that the voltage of the battery is labeled. Nearly all batteries that you will use deliver approximately 1.5 volts. The chemical reaction inside the battery always delivers the same amount of voltage regardless of the size of the battery. A big battery delivers the voltage for a longer period of time, but it delivers the same voltage as a smaller battery.

The exception is the *transistor* or 9-volt battery. If the chemical reaction of a battery only delivers 1.5 volts, how does a transistor battery deliver 9 volts? Open one up to find out.

Inside a 9-volt battery you will discover six battery cells that each produces 1.5 volts. Added in series—where the positive end of one is connected to the negative end of its neighbor—the voltages of the six batteries add up to give 9 volts.

Notice the size of the battery cells inside the 9-volt battery. There are small. They don't last long. As useful as these batteries are, they are a very expensive way to purchase electrical power. The terminals are easy to grip with alligator clip leads, but they wear out quickly. In general we recommend using AA batteries in a battery case to save money.

Two motors connected in series

Two motors connected in parallel

Connecting two motors

When you build an electric car you will probably get the idea to make it go faster by adding a second motor. If you connect the motors in series—so one motor terminal is connected to the other motor's terminal and the other terminals on each motor connect to the battery—you will find that the motors spin half as fast as one did when it was by itself. Now the two motors are sharing the voltage supplied by the batteries and each gets half the voltage.

The other way to connect the motors is to connect one terminal of each motor together and then connect the remaining terminals of each motor together. Then connect one end to one terminal of the battery and the other to the remaining terminal of the battery. This arrangement is called a parallel circuit.

Positive and negative

Although the batteries are marked with a positive and negative terminal we haven't suggested which end connects to which motor terminal. Place a tiny bit of masking tape on a motor shaft and connect it to a battery or battery pack. Notice what direction the masking tape is spinning. Then disconnect the battery and switch the position of the connecting wires so the wire that was on the positive terminal of the battery is now on the negative terminal. Complete the circuit by connecting the second wire. Which way is the motor spinning now?

By reversing the position of the wires on the battery (or on the motor) you change the direction of electron flow through the motor. The current flows through the motor in the opposite direction and so the motor spins in the opposite direction.

Super capacitors

Batteries are not the only way to power an electrical motor. Batteries store energy in the form of chemicals packed inside. The energy is released when the battery is connected to a complete circuit so a chemical reaction can occur. This allows electrons to flow out of one battery terminal and back in the other terminal.

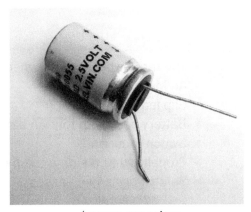

A super capacitor

Super capacitors hold electrical charge rather than generate charge from chemicals. In a super capacitor there is no chemical reaction. The reaction is a physical one in which electrons are packed very tightly together.

In general super capacitors cannot power an electrical motor as long as batteries can, but they are an effective replacement for many applications. And, they can be charged and discharged many thousands of times—essentially never wearing out.

Electric car with propeller drive

Use a fast-spinning motor to turn a propeller that moves a car. Be careful of the propeller as it has sharp edges that can cut your fingers!

Materials

- The three-wheel or four-wheel car you built in Project 1 or 2
- An inexpensive toy motor
- Clip leads (2 or 3)
- Battery pack with four AA batteries
- Switch (optional)
- Propeller that fits onto the motor shaft
- Safety goggles or glasses

An electric car model with propeller

Be safe

Wear safety goggles or glasses when using a propeller on a motor. Propellers can fly off.

Build the model

Mess around with the motor and the battery case, filled with batteries. Touch the terminals of the battery case to the terminals of the motor.

> > > > > > > > > > > > > > > > > Electric car with propeller drive

Does the motor spin? Put a piece of tape on the motor shaft to help you see that it is spinning. If it is not spinning check the battery case to see that the batteries are inserted in the correct direction. Look at the symbols on the battery case. The negative terminal (flat end) of each battery should push against a coiled spring. The positive terminal of each battery should be wedged in against a flat metal contact.

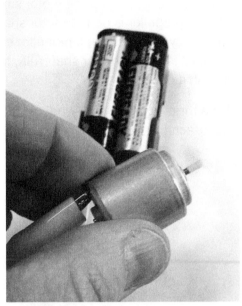

Connect the motor to barriers. Place a small piece of tape on the motor shaft so you can see the motor spinning.

Touch the battery case to the motor again, but reverse the connection. Change which terminal of the motor touches the positive terminal on the battery case. You should see the motor shaft spin in the opposite direction.

Use two clip leads to connect the motor and the battery case. Again the motor should spin. If it worked when you touched it to the battery case but does not work now, one of the two clip leads might be broken. Substitute a new clip lead and repeat the test.

This would be a good time to try adding a switch in the circuit. If you don't have a switch, make one. See the photos in the Materials Section in the beginning of this book. To add a switch you will need one more clip lead.

Connect all the components so electrons can flow from one battery case terminal to the switch to the motor and back to the other terminal on the battery case. Open and close the switch to see it stop and start the motor.

Push the propeller onto the motor shaft. To prevent injury ensure that the propeller is on the motor shaft as far as possible. At the high speeds toy motors spin, a propeller can fly off and hit you. If you aren't certain it is on the motor shaft fully, find someone with stronger hands to push it on.

Use a three-wheel or four-wheel model you built earlier. Where will you attach the motor with propeller? The propeller must be able to turn freely and not hit the ground or car frame. Do you want it at the back or front of the car? Where will you place the battery? Getting the car weight balanced with the battery and motor at opposite ends of the car would be best.

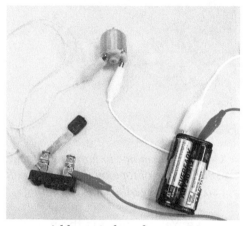

Add a switch to the circuit.

Use strips of cardboard or pieces of craft stick to make a motor mount. Cut a piece of cardboard about 15 cm long and 2 cm wide. Bend the ends of the strip so you can glue them to the car frame. Use a small drop of glue to attach the motor to the top of the mount. The glue will come off the motor later when you need it for another project.

A motor mount made of craft sticks

Use another drop of hot glue to hold the battery case in place. Connect the wires from the motor terminals to the battery terminals. At this point you probably do not know which way the motor will spin

> > > > > > > > > > > > > > > > Electric car with propeller drive

and which way the car will move. Make the electrical connections and hold your hand in front of the propeller to see what direction the air is blowing. Reverse the connections if needed. It is helpful to use different colored wires so you can remember which one attaches to the positive and negative terminals of the battery.

Put the car on the floor and turn the propeller by hand look to see if it hits the floor. When you connect the clip leads what happens?

Test the model

Place your electric car on a smooth floor and connect the battery to the motor. Does the car travel in a straight line? Does it go fast? It should shoot across the floor as fast as you can run to catch it.

Things to look for

If the car is slow, make sure that the wheels aren't rubbing against the frame or bearings. Also make sure that the motor is spinning fast. Listen to the motor for a high pitch sound. As the batteries wear down the pitch will lower and that will tell you it's time to change batteries.

Lay out a test track and mark the start and end with masking tape on the floor. A good distance for a track is 5 m (15 feet). Use a stopwatch (available on many smart phones) to measure how fast

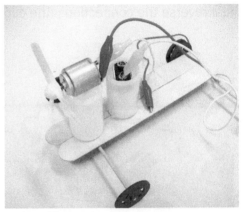

One film canister is the motor mount and a second canister holds a 9-volt battery.

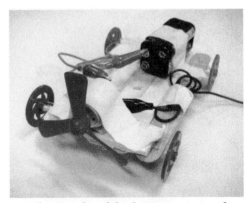

The weight of the battery case and motor are spread over the length of the craft stick car.

the car travels. Then ask yourself what single change could you make to get the car moving faster? Make the change and retest your car. Record the time for each test.

When you have the car moving fast, try reversing the leads to the motor to drive the car in the other direction. Time the car again to see which direction is faster. Does the car go faster with the motor at the back, pushing it forward? Or, does it go faster with the propeller in front pulling the car forward? Don't move the motor to do this test: just reverse the connections (the clip leads) to the battery.

Try this

Going straight is good, but how would it be to have your car travel in a circle? Racecars travel in loop circuits so try your car in a circle. Make the steering adjustable so you can change the size of the circle. See Project 3 for details on how to make the steering adjustable.

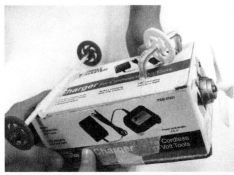
Two girls built this car. It completed 68 circles at a Seattle workshop before hitting a wall.

When we did this activity in Lulea, Sweden several cars raced around in a circle. One car completed 88 laps of the circle before hitting a wall and stopping. What is the largest number of complete circles your car can go (without you touching it)?

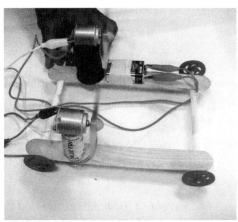
Two motors powered by one battery

What else could you do to make your car go faster? Would two motors speed it up? Can you connect two motors to one set of batteries?

How will you connect the additional motor to the battery case? There are two ways: series and parallel. If you aren't sure how to make these connections see the "Electricity primer" chapter.

After making a change to your car, time it on your test track. Record the time in seconds so you will remember which model is faster.

In the first you rolled a car down an inclined ramp. Do you still have the ramp? How far can your electric car go up the ramp? Does it travel higher when you let it build up to full speed on flat ground before it encounters the ramp?

Science concepts

If you measure the length of your test track and you record the time it takes your car to travel that distance you can calculate its speed. If the track is 5 m long and if it takes your car 2 seconds to travel that distance, the speed would be 2.5 m/sec. In an hour your car would travel 9 kilometers. Of course the batteries will wear out before the hour is up. How fast does a car or bicycle travel?

Three-wheel propeller car

This model moves in one direction by blowing air in the other direction. Feel the blast of air coming from the propeller and watch which way the car moves. How would the car move if you placed a sail on the car directly behind the motor? If you're not sure, try it.

Science project

This model offers many possibilities for science projects. You could measure the speed of the car with different numbers of batteries powering the motor. Battery holders for AA batteries are made to hold 1, 2, or 4 batteries. You could get a set of battery holders and mix them in series to deliver 1.5 volts, 3 volts, 4.5volts, 6 volts,

Cars Models that Zoom >

Battery holders for 2 and 4 AA batteries

A three-propeller model

7.5 volts, 9 volts, or 10.5 volts. Assuming the batteries don't weaken during your testing you would be able to gather 7 data pairs (voltage supplied by batteries and speed of the car) to make a nice graph. The slope of line would tell a story you could interpret.

At the speeds this car travels how important is aerodynamic shape? Would your car go faster if you streamlined the design? 9 kilometers per hour is the pace of a very fast walker or a very slow bicycle rider. If you are walking quickly or riding a bicycle slowly, are you concerned with wind resistance? At what speed does wind resistance become important?

At the speeds a car travels even a little increase in speed causes a big increase in drag or air resistance. The drag increases at a rate that is the speed multiplied by the speed or speed squared.

Electric car with direct drive

How could you make the car move without using a propeller? Could you attach a wheel directly to the motor shaft?

Materials

- The model car you made in Project 2
- A toy motor
- A battery case with batteries
- Clip leads (2)

Build the model

The most difficult part of this project is getting the wheel to fit the motor shaft. In most cases the center opening in the wheel is much larger than the shaft of the electric motor. How can you make these two different sizes of parts fit?

You need to buy or build a bushing. Bushings connect parts that don't fit by themselves.

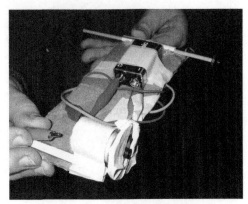

An electric car with direct drive

Check with the company that sells motors to see if they have a package of bushings. To make one will require a bit of creativity.

Bushings: A black rubber bushing is on the motor shaft (left). A yellow bushing is in the center of the wheel and a red bushing sits on top of the wheel.

Wrap tape around the motor shaft so the wheel will fit snugly on the shaft.

Insert a piece of wood dowel into the wheel and make a small hole for the motor shaft to fit into.

One way to get the shaft to fit snugly into the wheel opening is to wrap the shaft with a few turns of masking tape. Each turn of tape will take up a bit more room and at some point the wheel will fit snugly.

Another approach is to insert a piece of dowel into the wheel hole. Use the same size dowel you use for axles. Then drill a hole in the center of the dowel that will fit the motor shaft. Put a drop of glue into the hole and force the shaft in. Hold it in place until the glue dries.

Neither of these approaches is perfect. You will notice that the wheel isn't perfectly centered on the motor shaft. It will be close, however, and will probably work fine.

Where will you place the motor and wheel? The obvious choice would be to use it for the single wheel of a three-wheel car. Notice that when you attach the motor to the frame the wheel will be at a different height above the floor than the other wheels. But that doesn't matter.

> > > > > > > > > > > > > > > > > > *Electric car with direct drive*

Test the model

Connect the motor to the battery pack to see how fast the car travels. Does the drive wheel—the wheel that is attached to the motor—slip on the floor? If it does, how could you increase the friction of the wheel to prevent slippage?

Direct drive

Does the car travel in a straight line? If not, realign the drive wheel so it goes straight.

Use the test track you used to test the propeller car. How does the speed of the direct drive car compare to the propeller car?

Try reversing connections to the battery so the car travels in the other direction. Is the car faster when it goes front-wheel drive or rear-wheel drive? Often front-wheel drive cars steer straighter—especially on slippery surfaces—which is why many cars built for driving in ice and snow have front wheel drive.

Try this

Toy motors spin very fast. Some rotate as fast as 17, 000 RPMs, which means it makes 283 revolutions each second. That is too fast for this small car. Using lower voltages (using two AA batteries instead of four) would slow the motor, but might not provide enough power to drive the car.

A slower motor might work better for a direct drive car. Where do you get slower DC motors that are inexpensive? In general the slower a DC motor spins the more expensive it is. You can search catalogues of science supplies stores or can find them in old appliances and broken toys. Ask friends to save appliances that have motorized parts so you can take the appliances apart to get any motors.

When you find a motor inside an appliance, take it out carefully, leaving the electrical wires as long as possible. Test the motor by connecting it to a battery pack or 9-volt battery. Some motors will have markings on them that you can use to search online for the specifications. The specifications will show the lowest voltage the motor can operate on, the size of the motor shaft, and the speed of rotation.

Some of the motors you find in appliances will have more than two wires. Three wires identify a motor as a servo motor. Four or more wires tell you it's a stepper motor. Both require a microprocessor to operate so they aren't useful for these cars.

Try making the car drive in a large circle. To do this turn the drive motor and wheel so they are no longer in line with the other wheels. A small angle will cause a reasonable turn angle. A larger angle will cause the car to spin out. It will go forward for a short distance and then spin wildly in a small circle; or it might just spin in a circle and not move forward at all.

Science project

This model is very sensitive to the friction of the floor. A slippery floor will cause the model to lose traction. To illustrate the friction of different surfaces you could measure the speed of the car on the surfaces or observe the car's behavior on different surfaces.

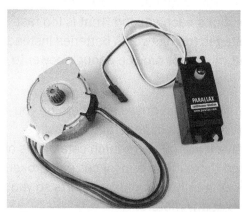

Stepper motor (left) and servo motor

Or, you could change the angle of the motor and wheel and observe how the car behaves. With the drive wheel aligned with the other wheels the car should go straight. When the drive wheel is at a small angle, the car will drive in a circle. As the angle gets larger, the circle the car drives will get smaller. At some angle the car will spin around and go nowhere.

> > > > > > > > > > > > > > > > > *Electric car with direct drive*

Science concepts

Why don't cars have direct drive? Engines have a limited operating range of shaft speeds. For gasoline engines to perform efficiently they have to operate from about 2,000 RPMs to 5,000 RPMs. But the wheels need to turn very slowly when the car is starting from a stop and they need to spin quickly when driving at highway speeds. To accommodate the range in speeds the wheels need, the motor's speed is reduced (or speeded up in overdrive) with gears.

⑫
Electric car with gear drive

Most automobiles use a gear transmission. Gears in a transmission convert the rotation of the motor or engine (a device that converts chemical energy into kinetic energy) to the wheels or axles. Usually they slow the speed of rotation as the motor spins faster than the wheels need to turn for safe driving. But they can also speed up the rate of rotation by using a different arrangement of the sizes of the gears. A small gear on the motor shaft that drives a larger gear will reduce the speed of rotation. A large gear on the motor shaft that drives a smaller gear attached to a wheel or axle will rotate the wheels faster.

Materials

- The model you made in Project 1
- An electric motor
- A small gear that fits onto the motor shaft
- A larger gear or a gear wheel
- Clip leads
- Battery case with batteries
- Balloon

Build the model

Slide the small gear onto the motor shaft. Hold the motor with the gear next to a geared wheel or large gear to see how the two will fit together. When the motor is spinning it will tend to push away from the gear wheel or large gear. So part of the design problem is to keep the two gears engaged (the one on the motor shaft and the one connected to the wheels). If they slip you will hear the teeth of the two gears grind against each other, but not engage.

Also, you will see the motor spin and car barely move or not move at all.

Gear wheel driven by gear on motor shaft

If you are using wheels with gear teeth, mount them on one of the axles replacing the other (non-gear) wheels. Hold the motor/gear on top of the car frame so its teeth mesh with the teeth in the geared wheel.

How will you keep the motor from sliding away from the geared wheel? Use a drop of glue to hold the motor in place on the car frame and hold the motor until the glue hardens. Put some glue on a short piece of dowel and push the glued dowel under the motor to prevent it from slipping away from the geared wheel.

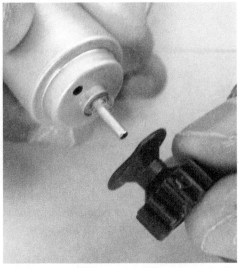

Slide a small gear onto the motor shaft.

When the glue dries you are ready to test.

Do you care that the motor will turn only one wheel? Probably not since the two wheels are both mounted on the same axle.

Gear on motor shaft drives a gear wheel

Cars Models that Zoom

Gear on axle driven by gear on motor shaft

A piece of dowel is glued between the motor and car frame to prevent the motor from moving away from the gear wheel.

Test the model

Connect a battery or battery pack to the motor. Are the drive wheels spinning? Can you hear the gears slipping? Hopefully not. Gently lower the model to the floor to see how it works.

Things to look for

Does the car move? When you put it on the floor does the motor stop spinning? The motor will stop when the load on it is too great. The car might be too heavy or the gears might fit together too tightly for the motor to spin. Realign the gearing so the gear wheel turns more easily.

Don't leave the power on the motor when it is unable to spin. Prolonged stalling (stopping it from turning) of a motor will damage it.

Do the gears slip? The sound will tell you if they are slipping. Try using a rubber band to pull the motor closer to the gear wheel without putting too much tension on the motor. Once you have achieved the right position and tension on the motor/gear the car will go flying across the floor.

Test this model on your test track. Have a friend time the car while you operate it. Record the time. Is this faster than the electric-propeller car?

> > > > > > > > > > > > > > > > > > *Electric car with gear drive*

If your model has rear-wheel drive, convert it to front wheel drive. Reverse the connections to the battery. Front wheel drive will probably provide a straighter course, but is it faster? Measure the time it takes for the car to travel the distance of your test track to find out.

The drive gear wheel is supported by the two craft sticks.

Do the wheels slip on the floor? Try adding tires. Cut off the neck of a balloon and stretch it over a drive wheel to give better traction.

Try this

If one motor drives the car super fast, what would two motors do? You will have to ensure that they are both spinning in the correct direction: one will spin clockwise and the other counterclockwise. If they both

Masking tape helps hold the motor in place.

are attached to the same axle and oppose each other's spinning, the car will not move.

Can you find other gears to use? As you take old appliances and toys apart always look for gears. By using different combinations of gears you can get the car to go faster or slower.

Science concepts

If two engaged gears have the same size same number of teeth, they both spin at the same rate. The same is true for two pulleys: if they are the same size they will spin at the same rate.

Add tires to the wheels by cutting off the necks of balloons and stretching them over the wheels.

A smaller pulley or gear on the drive shaft turns a larger pulley or gear at a slower rate.

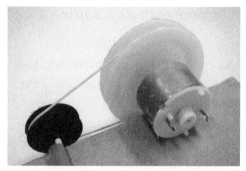

A large pulley or gear on the motor shaft causes a smaller pulley or gear to spin at a faster rate.

If the gear on the motor has half the number of teeth as the gear wheel, the wheel will turn at half the speed of the motor. It will take two turns of the small gear to turn the larger gear once.

If the size ratio is reversed so is the result. If the larger gear is on the motor shaft, it will spin one time while driving the smaller gear wheel twice.

While thinking about gears, go look at a bicycle that has gears. Many bikes have both a front set of gears and a rear set of gears. Turn the bike upside down so it rests on the seat and handlebars. Turn the crank and watch as you change the gears. Big gears driving little gears increase the speed for downhill, high speed biking. Little gears turning big gears help you climb steep hills. Gear ratio is the number of teeth on one gear divided by the number of teeth on the gear it is driving. If the front gear on a bike has 48 teeth and the rear gear has 12 teeth, the gear ratio is 4:1.

Electric car with belt drive

Most of the machines in factories and many trucks used to employ belts or chains to deliver power from an engine (or from a stream-powered water wheel) to the machine or wheels. Even today snowmobiles and some motorcycles use belt drives. For this model you can use a rubber band for the belt.

Materials

- The car you built in Project 1
- An electric motor
- A pulley that fits on the motor shaft
- A rubber band
- Clip leads
- Batteries in a battery case

Build the model

Use a rubber band to carry the energy from the spinning motor shaft to the drive axle. Look at your model to decide where to have the rubber band turn the axle. The rubber band should not rub against any other part of the car (the frame or bearing), so you might have to cut away part of the car frame or the axle

An electric car with belt drive

A pulley mounted on the motor shaft holds a rubber band that carries power to a pulley on the axle.

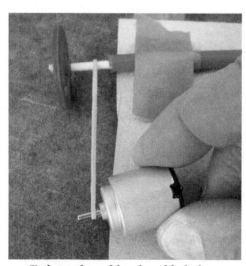

To keep the rubber band belt from sliding off the motor shaft, angle the motor slightly away from the axle.

bearing. Or, you might use a longer axle so it sticks out to one side to give enough room for the rubber band to hold.

One problem is how to keep the rubber band from sliding off the end of the motor shaft. If you have a pulley that fits snugly on the motor shaft, use that. The rubber band fits into the center of the pulley and hopefully will not slide off.

With or without a pulley you can angle the motor slightly away from the axle so when the rubber band slides on the shaft, it moves towards the motor and not towards the end of the shaft. If you make this angle too large the rubber band will rub against the motor and slow the car.

Before gluing the motor in place test the position by connecting the motor to a battery pack and seeing if the rubber band turns the drive axle. When you have just enough tension on the rubber band, move the motor a bit farther away from the axle to provide a bit more tension. You will need this additional tension when you put the car on the ground because the

> > > > > > > > > > > > > > > > > > > *Electric car with belt drive*

additional drag on the wheels will otherwise cause the rubber band to slip.

To prevent the motor from slipping toward the axle under the tension of the rubber band, reinforce the motor mounting. Cut a short piece of craft stick or dowel, cover it with hot glue, and insert it where the motor is glued to the frame. This will help hold the motor in place.

A gear and pulley is mounted on the motor shaft. A dowel is glued beneath the motor to hold it in place.

Use glue sparingly when gluing the motor the first time. You will probably have to reposition the motor after testing it.

Test the model

Time the car on your test track. Does it go as fast as the gear drive car? How can you get the car to go faster?

A dowel wedged between the motor and car frame keeps the motor from moving.

If the pulley on the motor shaft were larger, it would spin the axle faster.

Things to look for

See if the tension of the rubber band pulls the motor towards the axle. Check if the rubber band is so taut that the wheels can't turn or so loose that it slides easily over the axle.

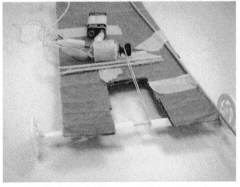

This model has a rubber band that pulls on the axle directly.

Cars Models that Zoom >

Normal arrangement of a belt makes the pulley rotate in the same direction as the motor shaft spins.

The belt is twisted so the pulley spins in the opposite direction as the drive shaft spins.

Try this

What would happen if you crossed the rubber band into a figure 8? You might find that this additional tension and rubbing stops the motor from turning. But if the motor can still turn, the drive wheels will go in the opposite direction.

Try this, too

Can you make a machine out of a shoebox? Mount the motor on top of the box and put an axle through the two end pieces of the box. Drive the axle with a rubber band that is turned by the motor. What you can make your machine do?

Solar-powered car

Here is a car that requires no batteries to run. It will move all day long...as long as the sun is shining.

Materials

- The car model you made for the electric gear drive or belt drive car
- One or two solar cells
- Low voltage or solar motor
- Lots of sunlight
- Super capacitor (optional)
- Voltmeter (optional)

This model works best with a low voltage or solar motor in place of the toy motor you've used in previous models. The toy motor can work, but will require more sunlight and additional solar panels. If you can get a solar motor or motor that operates on lower voltage, try that first.

Science concepts

Solar cells convert sunlight into electricity. Although they are not very efficient, they are efficient enough to provide the power to drive a model car.

Three solar cells are connected in series to power this model.

The motor turns a rubber band belt that turns the axle.

Two solar cells are connected in series to drive this motor.

As photons or light particles hit the material (semiconductors) inside a solar cell they add energy to the electrons in the material. The energy from photons causes the electrons to move and this provides the current for electric power. (This is a simplified explanation. Go online to find a more detailed description).

Use the solar cell(s) just like a battery. Connect a solar cell to the motor with two clip leads. Orient the cell so it directly faces the sun. Does the motor shaft turn? If not and if the sun is shining brightly, try connecting two solar cells in series. Clip the positive terminal of one cell (marked "+" or having a red wire) to the negative terminal (marked "-" or having a black wire) of the second one.

Connect the motor to the two terminals not yet connected. To help hold the two solar cells, motor, and clip leads, tape two craft sticks or pieces of cardboard to the back of the solar cells. Make sure both cells directly face the sun. If the motor still doesn't spin, try a lower voltage motor.

Experiment with the solar cells and motor. Pretend that your hand is a cloud partially blocking sunlight from the cell(s). What happens to the speed of the motor? If you completely cover one of the solar cells, what happens? If one cell is kept in the dark, it resists the flow of electrons and the motor won't spin even if the other cell has bright sunlight.

What happens as you change the angle of the solar cells so they do receive less direct sunlight? What can you say about the importance of orientation of solar panels towards the sun?

Playing around with the solar cells will give you design ideas for how to build a solar car. Even better is to measure the voltage output from the solar cell using a voltmeter. Science supply stores and catalogs sell inexpensive voltmeters that work well.

Use clip leads to connect the terminals of the solar cell to the probes of the voltmeter. Make sure the voltmeter is set to read DC volts of less than 10 volts. On most meters you twist a large dial to select what you want to measure. Look at the reading as you move the solar cell. Reverse the connection of the leads to the voltmeter. The readings should now be negative volts (if it showed positive voltage before).

This car uses three solar cells connected in series.

Use a volt meter or multi-meter to measure the electrical output of a solar cell.

Build the model

If you have a low voltage motor or solar motor, replace the toy motor you used for the gear drive car or belt drive car. The lower voltage motor will probably have a different size (it will be larger) so it will require you to make some adjustments to make it fit.

Solar cells will work best if they directly face the sun with nothing blocking sunlight. So the best place to locate them is on or above the car frame. If you use two or more solar cells, build a lightweight structure that will hold the cells and which you can mount on the car frame. Cut some strips of cardboard to make this roof for your car.

Connect the solar cell(s) to the motor with the clip leads as you would connect a battery case. With the cells facing the sun, the wheels should spin. Tape the wire clip leads to keep them out of the way.

If nothing happens and if you are using two or more cells check to see that you have wired the solar cells correctly. When wiring multiple solar cells in series each positive terminal should be connected to the negative terminal of the adjacent cell, except that the positive terminal of one cell goes to one side of the motor and the negative terminal of another cell goes to the other side of the motor.

If the wiring is correct or if you are using only one cell try changing the clip leads. One of them might be broken.

If you are inside you can try to use artificial lighting. However you will get better results outdoors with direct sunlight.

Two solar cells provide power to the geared motor on the car frame.

> *Solar-powered car*

Test the model

Set up a test track of a few meters length and mark the start and stop with masking tape. Put your solar-powered car on the start line and see how it moves. If it is going well, return it to the start line so you can time it on the test track.

A single solar panel car

What would make the car go faster? What is limiting its speed? If you have used just one solar cell can you add a second one? Or, add a third solar cell? Can you remove any weight from the car?

Look at the angle of the sun. How high is the sun above the horizon? Are your solar cells angled so they are perpendicular to rays of sunlight? If the sun is south of you and your cells are facing the north, the cells won't get much energy from the sun.

Would two solar cells drive two motors faster than two solar cells could drive one motor? If you think so, give this a try. Time each test and record the time.

Try this

Would a belt drive car work better with solar cells than a gear drive? Could you get a solar-powered motor to spin fast enough to the drive the car with a propeller?

Try this: night driving

Could you drive your solar-powered car at night? You could if you had stored enough solar energy. One way to store energy from solar cells is to connect the solar cells to a super capacitor. These devices store tremendous amounts of energy in a small space. You could connect a super capacitor to a solar cell or to two solar cells connected in

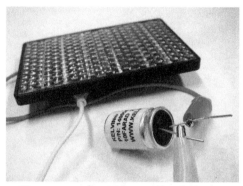

Use one or two solar cells to charge a super capacitor that can power the car at night.

series and let it charge for 10 minutes. In bright sunlight that will be enough time to fully charge the capacitor.

To connect a super capacitor to a solar cell, first look at the capacitor. It has one longer leg that is marked with ++++ signs. Connect this leg to the terminal on the solar cell that is marked +. Connect the shorter leg (-) of the capacitor to the negative (-) terminal of the solar cell. Orient the solar cell so it will intercept the most sunlight and let it collect energy for a few minutes.

If the sunlight is very intense where you are, the plastic of the solar cell might melt. This certainly demonstrates the power of the sun, but also can destroy the cell so watch the cell to make sure it doesn't get too hot.

If you have a voltmeter available touch the probes to the two legs of the capacitor. Turn the dial on the face of the voltmeter to measure DC volts with a setting less than 10 volts.

When you first put the solar cell in sunlight the voltmeter should show that there is no voltage in the capacitor. But within seconds you should see the voltage rise. When the voltage reaches the capacity (marked on the side of the capacitor), it is fully charged and ready to use. Even before this point the capacitor may have enough stored energy to drive the car.

Keep track of how long you charge the capacitor in sunlight. When you use the capacitor to drive the car also note how long the car moves. Compare how many minutes are required to charge the capacitor and how many minutes of driving it delivers. Are these about the same?

If you have two or more capacitors you could connect them in series to get a higher voltage and have the motor spinning faster. Or you could connect them in parallel (positive legs connected and negative legs connected) to drive the motor at the same speed, but for twice as long. The capacitors act like batteries would and can be connected in the same way.

Be careful not to touch the two legs of a capacitor together. This will cause it to discharge instantly and possibly be damaged. Also be careful that the two legs do not touch any metal or wire which would let them discharge.

Science concepts

You may have seen solar panels on the rooftops of homes or businesses. They collect solar energy and convert it to electricity. The individual cells each generate a small voltage just like the solar cells you are using. To get the higher voltages needed for household appliances, cells in rows are connected in series. To get higher current, the rows are connected in parallel. The direct current energy can be stored in batteries.

Solar cells generate direct current and homes and offices are wired for alternating current. To switch from direct current to alternating current, the solar system needs a device called an inverter. Once the energy is inverted to alternative current it can be used in the home wiring system and can be transmitted back to the electric power companies on the same wires that bring electricity into the house. A home with a good solar system and lots of sunlight can sell the electricity it generates instead of buying electricity from the power company.

A super capacitor is providing power for this model. The capacitor was charged with two solar cells.

Jittering car

Can you envision a car that doesn't have wheels? How would it move? Here is a fun vehicle that shakes to move.

Materials

- Choose one of the following:
  - Cardboard
  - Plastic disposable cup
  - Plastic food container (washed!)
  - Craft sticks
- Marking pens (optional)
- Toy motor
- Clip leads (2)
- Safety goggles or glasses
- Battery pack and batteries or 9-volt battery
- Propeller that fits the motor shaft

A jittering car

Build the model

Have you seen a cell phone that vibrates when there is an incoming phone call? Inside the phone is a small motor that makes the vibration. On the motor shaft is a weight this is not symmetric: it is off center. Every rotation of the motor throws the phone one way and then the other way.

You can use this same technology to make a car jump across a table.

The difficult part in making this model is finding an off-center weight that will stay on the motor shaft. One design is to start with a propeller. A broken propeller, with one or two blades broken off, works great. Or, use a propeller and add a small weight to one of the blades. Or, just wrap tape around the end of one propeller blade. Or, push a 2 cm piece of hot glue stick onto the motor shaft, ensuring that it is not centered.

In place of a propeller you could use a gear or other part that fits snugly on the motor shaft. Use hot glue and tape to hold a small weight to this gear. Be very careful as the weight might fly off.

Be safe

Before trying this off-center weight on the motor, put on safety glasses. The weight or propeller might fly off and you don't want to damage your eyes.

Motors that have off-center weights on the shaft vibrate.

With safety glasses on and a weight affixed to one blade of a propeller that is mounted on the motor shaft, connect the motor to a battery pack. Feel how the motor jitters and jumps.

Use some hot glue to hold the motor to a piece of cardboard that will be the frame. You could use other stiff material

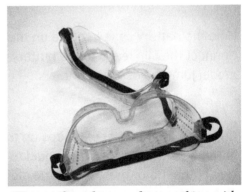
Wear safety glasses when working with propellers and off-center weights.

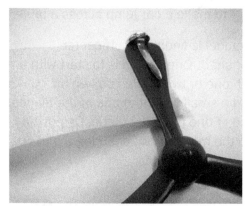

such as a small plastic drinking cup, disposable cup or a large craft stick. Position the motor so the propeller and weight extend far enough over the edge of the frame that they won't hit the frame.

Add a small paper fastener to the end of a propeller and secure it in place with a wrap of tape.

Mount the frame on three or four stiff legs to hold it above the floor. Craft sticks are easy to attach to the frame. If you are using a plastic cup you could hold the craft sticks to the cup with a couple of rubber bands or masking tape. To attach them to a piece of cardboard or disposable cup, use hot glue.

An alternative to using craft stick legs is to use stiff metal wire. A metal coat hanger cut into pieces will work.

Hold the battery case in place on the frame with a drop of hot glue. Connect the clip leads to the battery pack and motor. Watch the model jump across the floor.

Try this

Changing the number of blades on the propeller or changing the size of the weight will cause the model to change its motion. The more weight the motor must move, the slower it moves it. Larger weights give the car bigger jumps. Increasing the voltage (adding a second battery pack) will cause the motor to spin faster and the jumps to be wilder. Using a propeller with longer blades will also cause the motor to spin slower, but jump more wildly.

Try different combinations of weight and blade length to see which gives the most energetic motion. What would happen if you add a second motor with its own off-center weight?

Try this: electric art

Now can have your electric jittering car make some artwork for you. Replace the legs with marking pens. Or, tape the marking pens to the legs so the marking end of the pens will color a piece of paper on the floor. Tape the paper (a large sheet) to a smooth floor. Remove the caps from the pens and connect the motor. As the car jitters across the paper, it will leave either spots of color or solid lines of color.

The markings on the paper will change if you change the voltage, the length or number of legs, or the position or weight of the off-center weight.

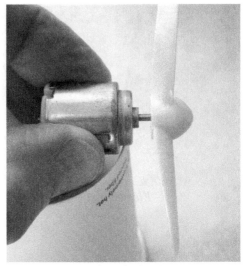

Position the motor so the propeller won't hit the cup or ground.

With marking pens as the legs the vibrations of the car draw lines across a piece of paper.

Cars Models that Zoom >

Electric art

Science concepts

Many appliances have motors with off-center weights. In addition to cell phones and pagers, electric toothbrushes and vibrating sanders use them. Electric football games use vibrators to shake the game field so the players move.

Hovercraft

Here is another model car that doesn't have wheels. It works by creating a thin layer of air beneath the frame that allows the hovercraft to glide over the floor.

This can be a challenging model to build so give yourself the time and materials to make several prototypes. This model only works on a smooth surface.

Materials

- ▶ Disposable paper bowls, or
- ▶ Disposable paper or polystyrene (plastic) dinner plates or bowls
- ▶ Toy motor
- ▶ Marking pen
- ▶ Single ply cardboard (cereal box)
- ▶ Propeller
- ▶ Clip leads (2)
- ▶ Battery (9 volt)

Build the model

The motor will perform two functions in this model. It will drive air underneath the frame to lift it up and allow it to slide over the ground. Instead of having wheels the hovercraft slides on a layer of air blown by the propeller. The motor also will propel the hovercraft forward.

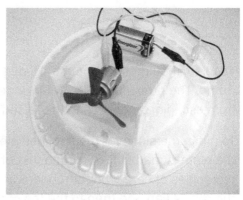

A hovercraft model

Materials to use for hovercrafts

We get one motor to do two different tasks by dividing the stream of air it creates into two parts. One part blows under the frame to lift it and the other part blows air parallel to the floor to push the hovercraft forward. The trick is to get the right balance between these two.

The frame can be one of several very light-weight materials, usually paper or polystyrene.

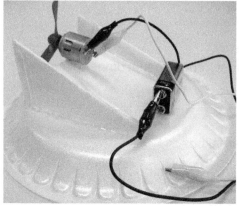

One propeller blows air both to lift and to push.

With a propeller on the motor shaft rest the motor on top of the frame. Use a marking pen to show the ends of the propeller on the top of the frame. You will cut a flap in the top of the frame that is as wide as the marks you just made.

With scissors or a knife cut three sides of a rectangle on the top of the frame to create a flap. Lift this flap up.

The flap should make about a 45-degree angle with the top of the frame. The propeller will blow air under the flap, but to keep the air from escaping add sides to the raised flap.

To make the sides for the flap, use pieces of polystyrene (the material the plate is made of) or cardboard cut from a cereal box. Cut two large rectangles of cereal box cardboard or polystyrene and hot glue one to the raised flap and then to the top of the frame.

> Hovercraft

Make sure you have filled all the spaces between the cereal box sides and the flap. When the glue has dried, cut the excess cereal box away. Repeat on the other side with the second piece of cardboard.

To mount the motor first cut a V or semicircle in the center of the flap. The cut should be wide enough for the motor to fit in. Then cover the cut surface with hot glue and hold the motor in place in the glue so the propeller will blow air into the raised flap.

Cut a flap out of the bottom of a polystyrene plate. Make the flap a bit wider than the propeller.

When the glue has dried, pick up the model. The hovercraft will work best if the weight on top (the motor and battery) is balanced—not all on one side. Decide where you can glue the battery to best balance the craft.

Use hot glue to fill the spaces between the plate and the sides.

We suggest you use a 9-volt battery for this model. The higher voltage of a 9 volt (over the 6 volts a battery pack would deliver) drives the motor faster and this model needs lots of air to be moved. And a 9-volt battery is lighter than a battery pack with four AA batteries. This model works best with the lightest weight possible.

Things to look for

With the model on the smooth floor connect the battery to the motor. Does the model move? If not, make sure the propeller is blowing air into the flap and not away from the flap. To reverse the flow of air, reverse the connections to the battery.

A 9-volt battery provides more power and less weight than a battery pack with AA batteries.

If the hovercraft still isn't moving, try it on a different floor covering. Also watch the hovercraft to see if it is turning about one point. This would indicate that the weight is off-center and that realigning the weight could allow it to move freely.

If you can't get the hovercraft to move, remove the battery and hold it in your hand. It should move easily with this weight removed. If it does, experiment to see where you can reposition the weight on the hovercraft to have it move.

To improve the motion of the hovercraft, try aiming the motor so more of the blown air goes into the flap. Or, reposition the motor lower on the flap so more air is caught. When it is working well it will fly across the floor.

Science concepts

The motor and battery are not powerful enough to lift the craft off the ground. You can try some experiments to convince yourself. Try building an electric helicopter and see if it lifts off. With a propeller on the motor shaft, hold the motor in one hand and have a friend connect it to a battery on the ground. Do you feel the motor lifting upward from your hand?

The motor and battery are strong enough to increase the air pressure underneath the frame (skirt) to raise the hovercraft just a bit. Of course as the sides of the hovercraft rise above the ground the air escapes and the pressure drops.

The weight of the hovercraft is spread over the area of the frame. If the total weight was 200 grams and the surface area was 200 square centimeters, the weight per area would be 1 gm/cm². This is the air pressure that the motor and propeller must generate to lift the hovercraft. In this example the air pressure required would be about 0.1% of atmospheric pressure at sea level.

A larger frame carrying the same weight would decrease the pressure required to lift the hovercraft. Or, if you decreased the weight of the hovercraft the pressure required would be less. Which option sounds better? Give it a try.

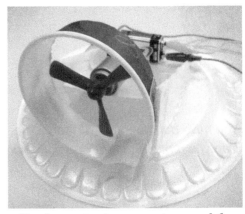

Putting a rim (paper cup) around the propeller can help increase the flow of air. This is called a ducted motor.

(17) Motorcycle

You have built models with four wheels, three wheels, and no wheels. Can you make one that rides on two wheels?

Materials

- Paper or polystyrene disposable dinner plates (4 or more)
- Dowels (1/4 inch diameter and 1/8 inch or about 6 mm and 3 mm diameter)
- Large diameter straws (these are milkshake straws with larger diameters than normal drinking straws)
- Motor
- Pencil
- Awl (sharp tool to make a hole)
- Clip leads (2)
- Battery pack with batteries or a 9-volt battery
- Propeller that fits the motor, or gear wheel and a gear that fits motor shaft

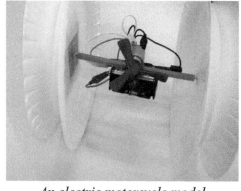

An electric motorcycle model

Build the model

Transform disposable dinner plates into large wheels. Are the plates strong enough to hold the weight of a motor and battery pack? If not, glue two plates together to become one stronger wheel. Build an identical second 2-plate wheel.

> *Motorcycle*

How will you arrange the wheels? It would be easier to have the wheels side by side, rather than having one in front of the other, but you can build either. We will show the side-by-side construction.

Find the center of the wheels you built. How do you find the center of a circle? One way is to find the balance point: support each wheel on the tip

Find the center of a plate by balancing it on a pencil.

of a pen or pencil. When the wheel is balanced, use the pen or pencil to mark the balance point. Do this on both wheels.

Use an awl or nail to make a hole through the center point of both wheels. Then push a pencil through to enlarge the hole.

Cut a piece of the larger dowel about 12 cm (5 inch) long to serve as the axle for your motorcycle. Push one end through the center hole in one of the wheels and glue it. Is the plate wheel strong enough to hold the glued axle firmly? If it is not strong enough add a support.

To make an axle support, cut a small square of cardboard from a box. Use the awl or nail to poke a hole through it and enlarge the hole with a pencil. Slide this onto the axle and glue or tape it to the plate. It doesn't matter if this support is on the inside or outside of the wheel.

With the axle firmly attached to one wheel, slide a 6 cm (2 inch) section of the large straw onto the axle. The motor and battery will attach to the straw. Add the second wheel to the axle, repeating the process you used to attach the first wheel. Make sure that the two wheels are parallel to each other.

Test your model

Roll your model across the floor. Does it roll easily? Do the wheels remain upright or do they wobble as the motorcycle moves? If the wheels wobble, strengthen their attachment to the axle. Cut more

Cars Models that Zoom

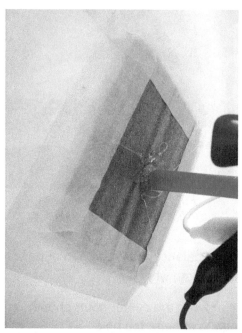

Make a support for the axle by gluing and taping a piece of cardboard to the plate wheels.

squares of cardboard, poke holes in the cardboard, and glue the pieces at the intersection of the axle and each wheel. To get a new piece of cardboard onto the inside of the wheel, cut a slot in the cardboard to the center hole so you can push it onto the axle before applying glue.

When the motorcycle wheel assembly rolls well add the motor.

Add propeller & motor

Use a small drop of glue to secure the battery or battery pack to the motor. Insert the propeller onto the motor shaft.

The battery and motor are glued to a large diameter straw. The wood axle is inside the straw.

Now attach this motor and battery assembly to the straw bearing that is riding on the wood axle. To be effective in moving the motorcycle the propeller should blow air directly opposite to the direction of travel. If the propeller blows air upwards or downwards, its force will be largely wasted.

Find the balance point of the motor and battery assembly and apply a drop of glue along this line. Hold the molten glue onto the straw bearing until the glue freezes.

When you release your grip the motor shaft should be parallel to the floor. If the shaft points upwards or downwards, pull off the motor and battery assembly and have another go at getting it placed at a more balanced position.

Coil a pair of clip leads and tape them together or use a rubber band to hold them. Leave just enough length to allow connections to the battery and motor. When making the electrical connections, do not let the propeller whack your fingers; it will soon be spinning fast.

Coil the electrical connectors so they don't drag on the ground.

The next thing you will hear are shouts of glee as the motorcycle races across the floor.

Test the model

Measure how long the motorcycle takes to travel the length of your test track. How does this time and speed compare to your earlier models?

Would the motorcycle go even faster if you used larger wheels? Could you make larger wheels out of cardboard boxes? Would it go faster if you reduced the weight? Can you find any material (weight) to remove that is not essential to the operation of the motorcycle?

After making each change, retest your motorcycle. By retesting after each change you learn how each change affects the speed. If you make five changes before retesting, you do not know which change made the biggest difference.

Build a new motorcycle

This model uses a gear drive. If you have wheels with gear teeth that fit the gear that goes onto the motor shaft, this is easy.

Mount one gear wheel in the inside center of each wheel. Glue the gear wheels to the paper or plastic dinner plates.

The gear wheels probably have a center opening that fits a 1/8 inch (3 mm) dowel. Glue a 10 cm (4 inch) long dowel in one of the gear wheels. Slide a 5 cm (2 inch) long piece of straw onto the dowel and then glue the other wheel to the dowel.

As in the earlier model the motor and battery will hang from the straw bearing. This time, however, the gear on the motor must mesh with one of the gear wheels. Before gluing the motor and battery to the straw check to see how well the gears will fit together. You might need to insert a small piece of cardboard to add space between the straw and motor/battery to get the gears to fit well.

Connect the two wires and watch it speed across the floor.

For a gear-drive motorcycle, glue the gear wheel to the plate and use a gear on the motor shaft.

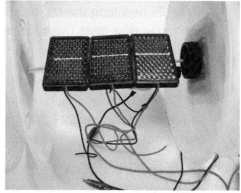

Solar-powered motorcycle

> *Motorcycle*

Science concepts

The propeller-driven motorcycle does not require friction with the floor to move. The motor is not directly turning the wheels—it is only pushing air. The wheels are free to turn or even skid across a smooth floor.

The gear-driven motorcycle, however, requires friction. Without friction the wheels would spin in place and the motorcycle would not move.

The large wheels make the motorcycle slow to start, but once going it is quite fast. If you put huge wheels, like bicycle wheels, the motorcycle wouldn't move at all. Would old CDs work as wheels?

Try this

Can you power a gear drive motorcycle with solar panels? You need strong sunlight and a smooth surface on the ground. Give it a try.

When you think of a motorcycle what image pops into your mind? The image is probably not like the models you just built.

Can you design and build a model that looks like the image of a motorcycle that is in your mind?

There are at least two ways to do this. One way is to make the two wheels wider so it is easier to balance on just two wheels.

To make wider wheels, glue two or more paper dinner plates together so the bottoms of the plates touch. This will make a wheel that is a couple of centimeters (1 inch) wide. You might need disposable plates made of stiffer material for this model.

Two gear wheels are glued together to provide better balance for this motorcycle.

Cars Models that Zoom >

A motorcycle with a tail!

The second design to consider is one that drags balance bars on the ground to keep the motorcycle upright. Consider these support bars as training wheels for a bicycle. These supports need to slide across the floor with the least amount of friction so they should be made of something that will slip easily.

In either model support the motor with propeller and battery between the front and back wheel. You could consider transferring power from the motor to one wheel by using a belt (rubber band).

Now you have mastered building car models with four (or more) wheels, three wheels, two wheels, and no wheels. Can you create a model that rides on one wheel?

(18) Cable car

Instead of supporting the weight of a car on the ground or on a thin layer of air above the ground (as in the hovercraft), support the weight on a string stretched across the room. Can you make a cable car that rides a string across a room?

Materials

- ► Cardboard
- ► Pulleys, gears, gear wheels
- ► Motor
- ► String
- ► Balloons
- ► Straw
- ► Craft sticks
- ► Dowels (1/4 inch and 1/8 inch or about 6 mm and 3 mm)

Build the model

Look for a location to stretch the string. The string needs to be pulled taut so you need to find solid anchors for the ends of the string. Electric appliances, chairs, and tables do not make good anchors. Water pipes make good anchors as do door handles on doors that won't open.

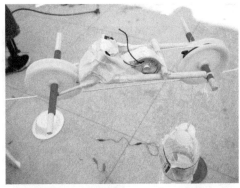

An electric motor drives this cable car.

Cars Models that Zoom >

Non-motorized cable car

Make a pulley by pushing two wheels together on the same axle.

Non-motorized cable car

Another way to anchor the string is to secure one end in a doorframe. Tie the string to a piece of wood large enough that it won't pull through the opening between the door and its frame. This only works for a door that won't be used during your experiments.

Supporting the weight of the car on a string makes the project easy because you don't have to worry about the direction of travel. Your car won't be able to turn sideways. However, using the string makes the project difficult because the cable car will have to be balanced.

Before making a motorized model, make a few non-motorized models that can slide along the string if one end is higher than the other. For this model you could use one or two pulleys and hang the cable car from them.

If you don't have a pulley or pulley wheel (a wheel with a groove around the center that the string fits), make one. Take two plastic wheels and push them together on a short piece of dowel. The two wheels mounted together form a pulley wheel.

> Cable car

Make a small car out of light cardboard. Attach the car to the pulley wheels or pulley so it is free to roll along the string.

Test the non-motorized model: with the cable car on the cable, lift one end of the cable (or depress one end) to see if the car will stay on the cable and if it will move downhill.

If the car falls off the string, the car might need additional weights held below the car. Or the problem could be that the pulley is not holding onto the cable. Watch the car to figure out what you need to change. When you have the car working well under the force of gravity, add a motor.

Add the first motor to make a jet power cable car. Add a balloon to the car. Can you get the car to travel the length of the string with one fully inflated balloon? Can you make it travel better by adding a second balloon?

For the balloon to push the car effectively the balloon exhaust must be oriented parallel to the string. If the balloon is venting upward, downward, or to one side most of the energy will be lost. To push the cable car forward the balloon must push in the opposite direction.

Non-motorized cable car

Tape the balloon to the straw.

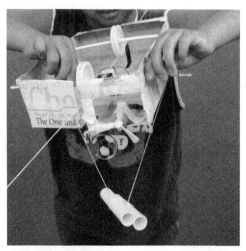

Propeller drive cable car

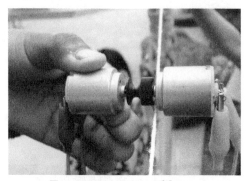

Two motors on a cable car makes it go fast.

Build an electric cable car

An easy way to motorize the cable car is to add an electric motor with a propeller. By now you have completed several projects that use a motor and propeller so this project will present few challenges.

Two potential problems are that the propeller might push sideways rather than push the car forward and that the cable car might start swinging from side to side. Watch what the cable car does and make adjustments so it will travel the length of the string.

Make the simplest electric cable car. This model eliminates all the parts that aren't absolutely required to make it travel. Slide a pulley onto the shaft of one motor, pushing the shaft on only half way. Push an identical motor into the other side of the pulley so you have two motors connected by one pulley.

The pulley will rest on the string, but of course it will be very difficult to get the two motors to balance. So use a 9-volt battery to both power the motors and to be the weight required to balance them on the string.

The trick here is to get the two motors to spin in opposite directions. One must spin clockwise and the other must spin counterclockwise.

> Cable car

Imagine the motor on the left side of the string. If it rotates clockwise it will push the cable car forward. The motor on the right side of the string must rotate counterclockwise to also push the cable car forward.

Use different colored clip leads so you can keep track of which wire goes to the positive and negative terminals. Find out which motor terminals to use and then use four clip leads to connect the motors to the 9-volt battery now hanging below.

Direct drive with string wrapped around axle pulley

The gear pulley should spin like mad. If it doesn't, check the wiring again. With the wiring correct remove two wires from the one motor so you can rest the pulley on the string. Reconnect the two wires and the cable car should fly across the string.

Try this

On a sunny day can you convert your cable car to one that is solar powered?

Can you make a switch so the cable car stops when it gets to the end of the string?

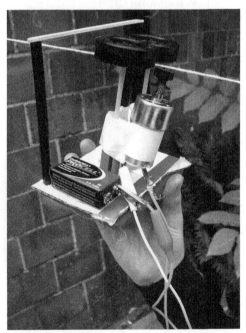

Gear-driven cable car. The string is held between the two gear wheels.

(19) Radio-control car

Building a radio-control (RC) circuit for a model car would be a major job, but here is a way to add radio-control to any of the electric models you have built so far. Use the remote control system from another toy.

Materials

- ▶ Electric motor car model from one of the previous projects
- ▶ Several discarded RC toys
- ▶ Masking tape
- ▶ Pen
- ▶ Wire cutters and wire strippers

Build the model

Find an old RC model car or boat. Look for one at thrift stores and ask friends if they have one they aren't using. You might need several used RC models to find one that has both the hand controller and the car circuit board that work.

If you can't find a used one, purchase an inexpensive model from a toy store or online.

Inside a radio control car

Carefully take the RC car apart. Inside you should find a circuit board, battery pack, and two motors. One motor drives the car forward and backward and the other motor controls steering. Some inexpensive RC cars only have one motor, but most have two.

Find the wires that go from the circuit board to the other parts. Label the wires by writing on a piece of masking tape and taping each piece to the wire it describes. For example, write Drive Motor, Steering, or Battery Pack wire. If you have a digital camera available take several photos that show the wiring. Once you have cut the wires it will be difficult to figure out which wire goes where unless you have a photo and tape markings on the wires.

Most RC cars have two sets of batteries. A battery case of four AA batteries provides power to the motors. You can substitute the battery pack that is already on your model for the case of AA batteries or use the RC cars battery case. A separate 9-volt battery provides power for the circuit board. The RC car needs both sets of batteries to operate.

Carefully remove the circuit board. To do this you will need to cut the electrical wires. Make sure that you cut them so the labels you just made are on wires connected to the circuit board. Cut the wires as far away from the circuit board as possible so you have the longest wires possible.

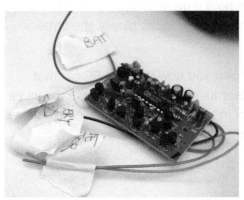

Circuit board removed from an RC car

Place the circuit board on the frame of the car you built. Connect the wires with the label Drive Motor to the motor on your car.

To connect two wires or a wire to a motor, strip 2 cm (3/4 inch) of insulation from the ends of the wires to expose the strands of metal wire.

You can wrap the exposed ends of two wires together or wrap the exposed end onto the motor terminals. Or, connect wires to the motor with clip leads if the wires from the circuit board won't reach the motor.

Strip the ends of the wires that went to the 9-volt battery so you can connect this to a 9-volt battery you mount on the car. Strip the ends of the two wires that went to the AA battery pack and connect these to your own AA battery pack. You are ready to test.

Use sharp scissors to remove the insulation from the ends of the wires

Make sure the 9-volt battery in the hand controller still has power. How can you test it? Many hand controllers have a small light that comes on when the power switch is turned on.

If the controller doesn't have this light (actually an LED), pull the battery out of the controller and hold it up to a motor to see if the motor spins. If the motor doesn't move or spins slowly, the battery needs replacing.

With some persistence on your part you now have an RC car that you've built. Can you make it steer as well? Or, can you use the wires that powered the steering motor in the old RC car to power lights or a motor that does something else?

Science projects

Once you have an RC car you can conduct a lot of different experiments. One simple one is to measure the range of the radio control system. In a large open area drive your car away from you until it no longer responds to the controls. Measure that distance.

Can you increase the range? You could increase the length of the antenna of the handheld controller. Attach a clip lead to the antenna to see if this increases or decreases the range. What else could you try? For some ideas, see my book, *Radio-Controlled Car Experiments*.

The author

Ed Sobey holds a Ph.D. in oceanography. He has conducted ocean research in Antarctica and throughout the Pacific Ocean and has written environmental analysis for Arctic and sub-Arctic regions of the United States. He has led expeditions to Southeast Alaska in sea kayaks and has participated in several SCUBA expeditions to collect marine specimens.

Ed directed five museums in the United States including the National Inventors Hall of Fame. He also founded the National Toy Hall of Fame. An active author, he has published 28 books.

Today Ed and his wife Barbara travel the world teaching for Semester at Sea and giving workshops for science teachers and science museum staff on inventing to learn science. The Fulbright Commission has given him grants to train teachers in Indonesia and Sweden. He has given workshops on five continents, in 25 countries. Visit his website: www.invention-center.com.

Some of Ed's other books suitable for young builders

Inventing Stuff, 1995, Dale Seymour Publications

Motor Boat Science—Hands-on Marine Science, 2013, Chicago Review Press

Unscrewed: Freeing Motors, Gears, Switches, Speakers, and More from Your Old Electronics, 2012, Chicago Review Press

The Way Toys Work, 2008, Chicago Review Press

Loco-Motion: Physics Models for the Classroom. 2005, Zephyr Press

Electric Motor Experiments, 2011, Enslow Publishers

Solar Cells and Renewable Energy Experiments, 2011, Enslow Publishers

Robot Experiments, 2011, Enslow Publishers

Radio-Controlled Car Experiments, 2011, Enslow Publishers

Rocket-powered Science, 2006, Good Year Books

CPSIA information can be obtained at www.ICGtesting.com
Printed in the USA
LVOW02s1515150114

369557LV00005B/20/P